Report on the Great Landslide at Frank, Alta., 1903

R. G. McConnell

R. W. Brock

Turtle Mountain and the great rock-slide at Frank, Alta., looking northward.
The cavity left by the breaking away of the central peak and the course of the slide can be distinctly seen. Under the slide-rock the terraced floor of the valley is discernible.
The uneven, billowed surface of the slide-rock is well shown and the heaped up rim along the edges of the slide can been seen at several points. In the foreground is a thin promontory of debris deflected from the main course of the slide. The town lies on the farther side of the slide in the direction of the gap between the mountains.

COPYRIGHT

Inquiries should be sent to:
Edmonton Geological Society
egs.publications@gmail.com

Printed in Canada

This Report was originally published in 1904 in Part VIII, Annual Report, 1903, Department of the Interior Dominion of Canada. It is reproduced with the permission of the Minister of Public Works and Government Services of Canada, and by courtesy of Natural Resources Canada, Geological Survey of Canada.

CANADIAN CATALOGUING IN PUBLICATION DATA

Edmonton Geological Society
Report on the Great Landslide at Frank, Alta., 1903.
ISBN 0-9697107-2-0

CONTRIBUTORS

Financial assistance is gratefully acknowledged from:

Alberta Energy Regulator

Alberta COMMUNITY DEVELOPMENT
Historic Sites and Cultural Facilities

AGS ALBERTA GEOLOGICAL SURVEY

This study edition has been prepared because the original document did not have such editorial conveniences as a Table of Contents, a List of Figures, a List of References and an Index.

The Table of Contents has been created by listing the headings throughout the paper. The major headings have been emphasized to clarify the structure of the Report. The original document contained only one class of heading. The List of Figures has been created by merging the 13 original Plates and the 2 Diagrams into a single list. So far as possible, figures have been placed in the text immediately after the first reference to them. The original description of the Figure has been added to the List of Figures.

The Map and Section in the Report were tipped into the original Extract from Part VIII of the 1903 Annual Report of the Department of the Interior. They were also issued loose in a separate pocket with other loose maps accompanying the 1903 Report of the Department of the Interior. A copy of the Map and Section is in the back cover pocket of the report.

References to Plates by superscripts in the text with corresponding footnotes at the bottoms of pages have been replaced by Figure numbers within the text. References to other technical publications again were by superscripts and footnotes; these have been replaced by much fuller references which recognize the modern reader's lack of familiarity with the early literature. Sufficient detail has been given to locate the publication in a large reference library or to order it through interlibrary loan. References by author and year in the text now refer to the List of References in the Report.

The titles of literature in German have been translated into English with the assistance of E. Eberhardt and S. Loew, (Zurich) and E. Consterdine (Uriage). Nigel Skermer's translation (1989) of Heim's Bergsturz und Menschleben discussed the English meaning of the first word of the title at some length because Skermer's choice, Landslide, did not fully convey the meaning of the term. While the events at Frank, Elm and Goldau were all landslides in the modern sense of the term, referring to them as "mountain collapses" here, now captures the dramatic tone of the German term.

No attempt has been made to modernize the geographic names or measurements used in the present text. Among recent changes in locality names are Crowsnest River for the middle fork of Old Man river and Crowsnest for Crow's Nest. As these names appear on the Map and Section they have been retained in the text and appear in the Index of Names and Localities that the text has generated. For similar reasons, names of geographic features, Old Man river, for instance, remain in lower case and the now usual, preceding definite articles have not been inserted. Measurements conform to the imperial system in use at the time.

Haanel's instructions, which have been added to McConnell and Brock's transmittal letter to create the Correspondence Section of the Report do not appear in the original Extract but are printed elsewhere in Part 8 of the Annual Report of the Department of the Interior. The instructions so clearly motivate the Report that it is incomplete without them. The portraits of McConnell and Brock (from the Geological Survey of Canada's collection) and their brief biographies are the only other additions to the text which was scanned from an original of the Report.

Further geological work by the Geological Survey of Canada (Norris 1993) and others (Cooper, 2000) has demonstrated that the rock displaced in the slide was from the east limb of the Turtle Mountain Anticline; the Report's cross section best describes the structure of Turtle Mountain north of the northern margin of the slide. The limestones are now placed largely in the Rundle Group, Mississippian Period rocks, formed about 350 million years ago. A field guide gives more details (Cruden, 2001).

Publication of this study edition of the Report has been supported by grants from the Alberta Historical Resources Foundation, the Canadian Geological Foundation, the Canada Landslide Loss Reduction Program of the Geological Survey of Canada and the

Canadian Society of Petroleum Geologists. S. Petaske skillfully processed the text and figures. This project was managed by Willem Langenberg of the Edmonton Geological Society.

The Editor, D.M. Cruden, may be notified of errors at the Department of Civil & Environmental Engineering, University of Alberta, Edmonton, T6G 2G7

References Quoted in the Preface

Cooper, M., Editor, 2000, Geological Highway Map of Alberta, Canadian Society of Petroleum Geologists, Calgary.

Cruden, D.M., 2001, A Guide to the Frank Slide, Technical Tour, 54th Canadian Geotechnical Conference, Calgary, 16 September, 12 p.

Norris, D.K., 1993, Geology and structure cross sections, Blairmore (West half), Alberta, Geological Survey of Canada, Map 1829A, Scale 1:50 000.

Skermer, N., 1989, Translator, Landslides and human lives, (A. Heim, 1932, Bergsturz und Menschleben, Zurich), Bitech Press, Vancouver, 195 p.

LIST OF FIGURES

OTTAWA, April 30, 1903

Messrs. McConnell and Brock –

You are to proceed at once to the scene of the disaster and investigate thoroughly the nature and causes of the catastrophe. If a landslide, determine as carefully as possible the condition of the strata before the avalanche, and the causes which led to the breaking away of the mountain mass, whether due to imperfect or careless mining operations, the explosion of fire damp, exceptionally bad ground, or similar causes, or whether the results were due to the destructive effects of atmospheric weathering or other causes.

Determine with especial care the present physical condition of the mountain, with a view of determining the likelihood of a recurrence of similar destructive phenomena.

Report as soon as possible, consistent with thorough examination, the results of your investigations.

DR. EUGENE HAANEL,
Superintendent of Mines,
Department of the Interior.

OTTAWA, June 12, 1903.

Sir,—In pursuance of the instructions of the Honourable the acting Minister of the Interior, transmitted through you, we have examined the Frank landslide, and beg to submit the following report thereon.

We have the honour to be, sir,

Your obedient servants,

R.G. McCONNELL

R.W. BROCK

Geological Survey of Canada.

Turtle mountain (Figure 1), the scene of the late disastrous rock avalanche, forms part of the most easterly or front range of the Rocky Mountains. It is a long, narrow, somewhat wedge-shaped ridge, surmounted by a number of rocky peaks, the highest of which have an elevation of rather more than 3,000 feet above the valley of Old Man river, or more than 7,100 feet above the sea. The range, of which Turtle mountain forms a part, runs in a N.N.W.—S.S.E. direction. It is pierced, north of Turtle mountain, by a narrow gap, through which Old Man river secures an exit from the mountains. Half a mile south-east of the gap, where the valley is broadened by the debouchment of Gold creek, and close to the foot of the mountain, nestles the town of Frank, an important coal mining centre on the Crow's Nest branch of the Canadian Pacific Railway. The summit of the Rockies and the continental divide lies 14 miles to the west.

Turtle mountain is a typical Rocky Mountain ridge. It is built of westerly dipping beds and its steep sides meet above in a sharp crest, with projecting rocky points. The eastern face of the ridge, the one which overhangs the town of Frank, possesses two main slopes. The lower or talus slope is 800 feet in height at Frank, and rises at an angle of about 30°. The upper or cliff slope is excessively steep, the angle of elevation throughout often exceeding 45° and rises in a succession of precipitous limestone cliffs, some of which appear almost vertical. The central peak, the one which broke away, is stated to have projected eastward beyond those that remain and probably rose above the steepest portion of the ridge (Figure 1). The western slope is more uniform and flatter than the eastern, the angle of elevation averaging about 33°.

Between the Gap and Frank, Turtle mountain rises directly from the valley, but opposite the south peak a wide spur, 1,200 feet in height, is thrown off to the eastward, which descends to the level of the valley in a succession of terrace-like steps (Figure 2).

Old Man river is a rapid stream, 40-50 feet in width. After issuing from the Gap, it flows southward along the base of Turtle mountain until deflected to the east by the spur from the south peak.

The valley of Old Man river, near Frank, consists of a flat about half a mile in width, trenched slightly by the beds of Old Man river and Gold creek, and broken in places by low sandstone ridges and knolls. A conspicuous hill of broken rock south of the railway track was formerly a rocky knoll 50 feet in height (Figure 3). The valley is bordered on both sides by terraces. A narrow, well-marked terrace, built of boulderclay and gravel, follows the river on the right from the Gap southward to near Frank. It is covered with debris or destroyed for some distance east of Frank, but appears again east of the slide, at a height of 150 feet above the river. The north bank is interrupted by two main terraces. The lower has a height of fifty feet above the railway flat east of the slide, This terrace separates into two smaller ones, before reaching the eastern end of the slide. The upper terrace has an uneven surface and its scarp is notched in places by subordinate terracing. It has a height above the valley of from 250 feet to 300 feet. The surface east of the terraces is contoured irregularly, but rises steadily eastward toward the Livingstone range.

The main points in the topography of the region are the precipitous slopes of Turtle mountain ridge, the flat-bottomed terraced valley of Old Man river at its base, and on the farther side a gradually rising surface, terminating in a high ridge.

Turtle mountain may be described generally as built of Upper Palaeozoic limestones above and Cretaceous shales and sandstones below. The northern part of the mountain consists entirely of limestone, but going southward, the underlying sandstones and shales rise gradually in the face of the ridge, and a mile south of Frank reach an elevation of about 1,300 feet above the valley of Old Man river. The dip of both series is westerly at high angles.

Figure 1
Valley of Old Man River and Turtle Mountain before the slide.
The left peak is the one which broke away. The right peak is the part of the mountain which still threatens the town.

Upper Palaeozoic

The limestones in the lower part of the mountain are evenly bedded, are very compact and as a rule are dark in a fresh fracture, but weather to a light gray, or yellowish colour. Ordinary and magnesian limestones are both present, and many of the beds are more or less bituminous. A band of dark shales, 25 feet in thickness, occurs about 800 feet above the base of the limestones, and is succeeded by shaly impure limestone, which in places has a strongly developed cleavage. These beds mark the base of the cliff portion of the mountain above Frank. The limestones in the upper portion of the mountain are more crystalline than those below, are purer and occur in heavy massive beds, separated by flaggy bands. A zone of conspicuously striped cherty limestone consisting of alternating layers of gray limestones and black chert from one to three inches in thickness, occurs some distance above the band of shales. The limestones outcropping between the crest of the mountain and the Cretaceous beds which overlie them on the west, were not examined.

Age of Limestones

Two corals, one collected about 1,500 feet above the base of the limestone, and the other from the slide rock, were submitted to Mr. Lawrence Lambe, assistant palaeontologist, who contributes the following note in regard to them:—

'*Diphyphyllum arundinaceum,* Billings. A well preserved specimen, with external characters and the internal structure clearly shown. This species occurs abundantly in the Corniferous limestone of Ontario. A loose specimen from Peace river, B.C., between Fossil Point and the Canyon on the Mountain of Rocks, was collected by Professor John Macoun, in 1875, (Devonian).'

'*Acrophyllum Oneidaense,* Billings. The second specimen probably belongs to this species. It is fragmentary and does not present the structure in a thoroughly satisfactory manner, but sufficient data are available for a specific determination. This species is also characteristic of the Corniferous formation, as developed in the province of Ontario.'

'The above two species clearly indicate a Devonian horizon.'

These fossils determine the position of a portion at least of the limestone outcropping on the eastern face of Turtle mountain, where the slide occurred. The prevalent westerly dip brings in higher beds on the west and it is probable that these extend up into the Carboniferous.

Cretaceous Beds

The Cretaceous beds along the eastern face of Turtle mountain consist mostly of alternating bands of hard, in places almost quartzitic, rusty coloured sandstones and dark, comparatively soft shales. The shales weather easily and are seldom exposed, while the harder sandstones often project above the surface in hummocks and long ridges. The beds include an occasional band of coarse pebbly conglomerate, some carbonaceous shales and some important coal seams, one from 13 to 18 feet in thickness. The dip of the beds is westerly at an angle of 82°, and the strike nearly north and south.

Structure of Mountains

The limestones which form the main mass of Turtle mountain range, and which are in part at least of Devonian age, dip steeply to the west at an average angle of about 50°. They are overlaid by Cretaceous sandstones and shales along the western base of the range, but they overlie beds belonging to the same series on the eastern. The peculiar relationship of the two formations can only be explained by faulting. The front ranges of the Rocky Mountains, wherever they have been closely examined, have been found to owe their origin largely to a great system of thrust faults. Near Bow river (McConnell, 1886), the Cambro-Silurian has been broken and thrust forward for miles over the Cretaceous and at the International boundary Mr. Bailey Willis (1902) has shown that the rocks down to and including a part of the Algonkian, have been similarly affected and now rest on Cretaceous strata. At Turtle mountain the thrust has not been so great, but has been extensive enough to bring the Devonian and possibly somewhat lower rocks over the Cretaceous. The fault plane near the surface dips steeply to the west, but the exact angle was not ascertained. (See section of Turtle mountain.)

A short, sharp double fold occurs about 800 feet above the base of the limestones, a second minor one about 200 feet lower down and some of the lower beds are badly crushed into angular, irregular shaped fragments, often with slickensided surfaces. With these exceptions the dip of the strata is remarkably regular.

The Cretaceous beds, on the other hand, have been completely overturned east of the fault, and now dip westward at angles of about 80°, and the whole formation for several miles to the east has been strongly folded and crumpled.

The limestones are traversed by several sets of jointing planes, intersecting each other at various angles. This structural feature is important from its possible bearing on the cause of the slide. Two strong sets intersecting at an angle of 97° are conspicuous at several points along the crest of the ridge, and their effect on the erosion of the mountain is plainly seen in the successions of salient and re-entrant angles which line vertically the face of the ridge.

Figure 2
Mouth of the Canadian American Coal and Coke Company's mine and slope of Turtle Mountain before the slide.
The coal seam runs from the tunnel up the slope to the point on the hillside marked A.

The fissures opened up by the slide follow the jointing planes, as a rule, but often break across from one set to another. The limestones are also cut by old fracture planes, probably formed during the faulting which produced the mountain. More or less movement must have occurred along some of these planes, as slickensided surfaces are common in the debris brought down by the slide.

At dawn, on April 29, 1903, a huge rock mass, nearly half a mile square and probably 400 to 500 feet thick in places, suddenly broke loose from the east face of Turtle mountain and precipitated itself with terrific violence into the valley beneath, overwhelming everything in its course. The great mass, urged forward by the momentum acquired in its descent, and broken into innumerable fragments, ploughed through the bed of Old Man river, and carrying both water and underlying sediments along with it, crossed the valley and hurled itself against and up the opposite terraced slopes to a height of 400 feet. Blocks of limestones and shale, mingled with mud, now cover the valley to a depth of from 3 to probably 150 feet, over an area of 1.03 square miles (See map, section and frontispiece).

The number of people killed by the slide is not known exactly, but it is given at about 70. The property destroyed includes the tipple and plant at the mouth of the Canadian American Coal and Coke Company's mine, the company's barn and seven cottages at the east end of the town of Frank, half a dozen outlying houses, with some shacks and camps, besides a considerable number of horses and cattle and a couple of ranches. The track of the Crow's Nest Railway was hopelessly buried for a distance of nearly 7,000 feet and the lower mile of the Frank and Grassy Mountain Railway met a similar fate. The people occupying the houses in the track of the slide were all swept away with it and destroyed, with the exception of a few near the edge of the slide, who escaped in some almost miraculous way, which they themselves cannot explain (Figures 3 and 4).

The slide occurred about 4.10 a.m., at a time when most of the inhabitants of the valley were asleep and before full daylight. The statements of the few eye-witnesses throw little light on the character of the slide, but the following notes obtained from them are not without interest:—

Karl Cornelianson was awakened by the noise. He rushed to the door of his house, which looked out over the first terrace flat and the base of the second. His first thought was that there had been an explosion at the mine and his first look was in that direction. Seeing nothing there; he glanced round to the terrace flat in time to see the rock debris hurl itself against the slope of the second terrace, and its momentum spent, fall back to the lower level. His impression was that an explosion had taken place directly in front of him. The edge of the slide was about a quarter of a mile from his door.

Mr. McLean, who kept a boarding house in Frank, was already up. Hearing the noise, he rushed to the door in time to see the slide rush by only a few feet in front of him. The passage of the slide was so instantaneous that he thought an eruption had taken place directly in front of him.

A freight train was shunting on the mine siding at the time of the disaster. They had taken some cars of coal from the tipple at the mouth of the tunnel and were slowly backing up for another load, when the engineer heard the rocks breaking away from the mountain above. He immediately changed to full speed ahead and ran out of danger. The conductor saw the men at the tipple become alarmed and start to run, but they were overtaken by the slide and perished. Immediately after everything was shrouded in a cloud of dust.

Mr. Warrington, who was sleeping in one of the cottages destroyed, was awakened by a noise which he thought was caused by hail. He jumped out of bed and then realized that it was something more serious, but before he had time to become alarmed the house began to rock, and the next thing he was conscious of was finding himself in the lee of some rocks, forty feet from where the house had stood. His bed was some twenty feet farther on. His thigh was broken and he was otherwise injured by small fragments of debris being forced into his body. He pulled himself out with his arms, and was trying to work his way to some children, whose cries he heard, when the first rescue party arrived.

Mrs. Ennis, whose house was also destroyed, was startled by the noise. She had time to waken her husband and get out of bed, but before they could rouse the children their house was demolished.

Figure 3
Valley of Old Man River and the Town of Frank before the slide, looking northward from sandstone knoll south of railway track.
To the left is the north spur of Turtle Mountain showing terrace at base, talus slope and a little of the cliff portion of the mountain. To the right is Bluff or Goat Mountain. To the right of the gap between them Crow's Nest Mountain is faintly discernible. Old Man River passes through the gap and flows at the base of Turtle Mountain. Gold Creek can be seen crossing the middle of the valley just below the first houses of the town. The barn at the left and the first row of cottages were buried under the slide.

Lester Ackroyd (Johnson), whose father and mother were killed, was not wakened. He found himself under the floor of the house and escaped through a small hole between it and the surface of the slide rock. He looked for his neighbours' houses, but finding them gone, waded the creek and proceeded to a friend's cottage. He was seriously wounded by a splinter which pierced his abdomen.

Many of the inhabitants of Frank, who were wakened by the noise, state that the thud of the rocks striking the valley bottom was distinctly felt, but that the shock differed entirely from an earthquake movement. The noise of the slide was described as resembling that of steam escaping under high pressure.

Nineteen men were working in the mine at the time of the slide. Of these seventeen escaped and two, who are supposed to have been at or outside the mouth of the tunnel, perished (Figures 2 and 5).

The statement of Chapman, one of the miners working at the top of No. 8 man-way, of what happened in the mine is as follows: He became aware that something was wrong at 4 a.m., according to his watch. The coal commenced to break and run down the man-way. The men working near by became alarmed and started to escape down the ladders. One of them, Mackenzie, was caught by the crushing in of a cross-cut, but Chapman and a companion succeeded in rescuing him before he was smothered. In coming down the man-way, they were considerably battered by falling coal, but not seriously injured. After reaching the gangway, they proceeded along it to the entrance, but found it blocked with debris. The lower tunnel was then tried, but proved to be full of water. After some discussion as to whether it was better to dig their way though the debris closing the mouth of the tunnel, or to send an upraise to the surface, they decided on the latter course. Volunteers had to go back to the face of the tunnel, 5,500 feet, in order to secure the necessary tools. It was 7 a.m. before they commenced the upraise and nearly 5 p.m. before they reached the surface. During the operation men went back into the mine for timber and Mackenzie tried to reach the surface through an air-way, but was turned back by gas. These trips back into the mine show that the main passage ways at least were not seriously injured.

Miners who were in the gangway at the time report that a blast of air rushed into the tunnel, blew the trap doors open and threw men and horses off their feet. Immediately afterward the mouth of the tunnel was sealed with debris.

Figure 4
Valley of Old Man River and the Town of Frank after the slide, looking northward from nearly the same point as in Figure 3.
The large boulders and mud flats are well shown. The C.N.P. Railway is being reconstructed across the slide.

Classification

Baltzer (1880) divides large land slips into two classes, the Bergsturz and the Bergrutsch. In the former the rocks break across the bedding planes and in the latter slip along them. The Turtle mountain slide is a typical example of a Bergsturz, as the break cuts across the bedding planes, almost at right angles.

Time and Rate of Movement

It is difficult to arrive at any definite conclusion in regard to the time occupied by the slide, as the estimates of eye witnesses range all the way from twenty seconds to two minutes. It is probable from the evidence obtainable, such as it is, that the time which elapsed between the first crash and complete rest did not exceed one hundred seconds and may have been somewhat less. The distance from the summit of the mountain to the end of the slide, measured horizontally, is 9,916 feet, and following the slope, about two and a half miles. No estimate of the rate at which the material travelled after it broke away can be given, but as those awakened by the roar had scarcely time to do more than to rise from their beds before all was over, it must have been extremely rapid. Heim (1882) estimates that the blocks in the great slide which occurred at Elm in 1881 travelled about a mile and a half in from 10 to at most 30 seconds. The material in this slide leaped from a projecting shelf right across the valley, and the movement on this account was probably more rapid than in the case of the Turtle mountain slide.

Character of Movement

The separated rock mass seems to have been shattered by impacts against the side of the mountain during its descent, and probably long before it reached the bottom, into myriads of fragments, some of which were doubtless flung far out into the valley. A shelf of rock in the basin of the slide seems to have hurled most of the material over the coal mine at the base of the mountain into the river bed, or beyond. The movement of the broken rock mass cannot be characterized as a slide in the ordinary sense of the word. The blocks must have travelled to their destination largely by a succession of great leaps or ricochets, probably accompanied by a certain amount of rolling and sliding. The character of the movement is clearly shown in the gradually lessening bounds ending in a short roll of a number of fragments, which were thrown forward beyond the main mass. The progress of these can be distinctly traced by the indentations made in the surface by the bounding rocks. While the movements of the individual fragments consisted of a succession of bounds from the surface and caroms from flying rocks, the movement of the mass, taken as a whole, suggests that of a viscous fluid. On the level flats the movement was onward, but with a tendency towards lateral dilation, but when terraces or other elevations were encountered, a portion of the material was deflected

Figure 5
Dominion Avenue, Frank.
The treeless portion of the mountain side represents a section of the slide. The cloud of dust at the top is caused by falling rock. Just above the houses on the right can be seen the crushed and crumpled beds near the base of the limestone series which weather out into pinnacles. The group of men coming up the road are the miners who have just escaped from the mine. In the wagon is one who was injured. The point of escape lies in the rock slide behind the last house on the left.

and flowed along the obstruction (See Map, back cover pocket). In some instances the rocks appear to have been flung back from steep terraced faces like spray, or fell back after climbing part way up (Figure 7). The outline of the slide consists of a succession of bays and promontories, most of which are largely due to the irregular topography of the valley. The total deflection in some cases amounted to fully 90°.

The cessation of the movement appears to have been remarkably sudden. At a number of points around the edge of the slide the rocks piled up into a high rim and then, the velocity acquired in their descent becoming exhausted, fell gently forward. (See frontispiece). This seems to be the explanation of the escape of a number of trees which still stand erect a few feet inside the slide. The trees are surrounded by rocks, but have apparently sustained little injury.

The Slide Rock

The slide rock consists mostly of angular fragments of limestone, ranging in size from grains up to great blocks forty feet in length. Large rocks are common everywhere, and in places, especially along the central portion of the slide, the greater part of the debris consists of fragments from three to twenty feet in diameter (Figure 6).

Dark shale fragments are abundant along a narrow belt, running parallel to the north-western edge of the slide, and are sparingly distributed elsewhere. Similar shales occur in the mountain, interbanded with the limestone. No rocks of Cretaceous age were found in the debris, and it is assumed that the slide did not cut down to them.

In portions of the slide the spaces between the rocks are filled with material resembling boulder clay and a number of small mud flats, separated by mounds and ridges of broken rock, occur east of Gold creek (Figure 7). The edge of the slide is usually fringed with more or less mud and the same material, mingled with small rocks, occurs in places spattered over the surface for some distance ahead of the slide. At some points water was squirted out from the slide, laden with trees, mud and a few boulders. The soft mud was derived largely from the bed of Old Man river, parts of which were scooped out, probably down to bed rock. The bed of Gold creek and the marshes in the road of the slide furnished a portion of the material and a further contribution was obtained from the fragments of the boulder clay terraces on the lower slopes of Turtle mountain.

Surface of Slide

The surface of the slide is singularly uneven (Figure 8). The rocks are heaped up into mounds and short interlacing ridges inclosing hollows, somewhat resembling a terminal moraine. In some places the lumpy condition of the surface suggests that the material travelled in waves, the waves retaining their form when movement ceased. Peculiar conical mounds, built of loose rocks up to three feet in diameter, were noticed at several points. The origin of these mounds is uncertain, but some of them may represent portions of ridges, which have been partly destroyed by flying rocks (Figures 9 and 10). The larger ridges are due in most cases to elevations in the old surface. A well marked depression extends across the slide, near

Figure 6
The slide-rock in detail.

the line of the old railway. It appears to be due to the protection of a rocky knoll to the south, which flung the greater part of the material across it. This depression is followed by the new grade of the Crow's Nest Pass Railway.

The slide rock usually terminates abruptly in a steep slope from six to thirty feet in height, and in places is heaped up at the edge into a prominent rim. The rimmed character of the edge is especially noticeable where the drive was straight onward. The edges of the promontories where the material spread or was deflected sideways are usually comparatively thin. (See frontispiece).

The mud flats mentioned previously as occurring east of Gold creek form a conspicuous feature of the surface. The soft semi-liquid mud was splashed out of the bottom of Old Man river by the falling rocks, and driven for some distance ahead of them. Afterwards it spread out between the debris piles into small level spaces. Diminutive cones with crater-like depressions in the centre, due to the escape of imprisoned air, are common features of the mud flats.

Changes in Topography

Changes in topography due to the slide are common. The valley is, of course, completely changed, as the former flats and even terraces are buried deeply under a mass of unassorted material. Old Man river is dammed at the lower part of the slide, and the channel above it occupied by a chain of small lakes with a depth of from 20 to 35 feet (Figure 11). The bottom of the lake opposite the tunnel is several feet lower than the old bed of the river. Gold creek has been deflected to the west and now follows the edge of the slide to its junction with Old Man river. A couple of small lakes are formed, where streams coming from the north reach the edge of the slide, and a shallow lake which formerly existed near the east corner of the slide has been filled up.

Figure 7
Looking eastward along the slide.
The foreground is a mud flat. Some of the large boulders are coated with mud or have it dripping down their sides. The largest boulder is 40 feet long. A high ridge of boulders forms the background. On the left of the edge of the slide the hillside can been seen against which the boulders washed. The uneven surface of the slide is well shown.

Some Minor Features

While the boulders as a rule are much bruised, a few in the slide just east of the river have slipped down to their present position with little or no bounding or rolling. One large boulder still retains on its upper surface a coating of moss and a fragment of a rotten log lying quite undisturbed. This must have broken away about the close of the slide. From its position and the character of the wood, it is evident that it came from the lower part of the northern limit of the slide.

The huge rock mass, falling with such frightful velocity, must have greatly disturbed the lower strata of air. The air must have been compressed under it and forced out as a sudden heavy gust of wind. The miners, as already mentioned, felt a strong blast in the gangway of the mine, but this was not noticed in the town above the slide. On the other hand, the southern edge of the slide for some distance away is strewn with twigs and branches of trees and other wreckage, evidencing a rush of air down the valley. The cause of this southerly escape of the air may have been the natural wind, which was blowing down the valley at the time, or may have been due to the northern part of the mass breaking away a few seconds in advance of the southern.

An avalanche just south of the edge of the slide is partially covered with debris. The heaviest deposits of dust are also south of the slide.

A quantity of dynamite was stored near Poupore and McVeigh's construction camp, situated at the junction of the Grassy Mountain Railway and the Crow's Nest Pass Railway, near the south edge of the slide. It was probably exploded during the slide, but the only evidence of the explosion is furnished by the boulders buried in the ground over an area extending at least 2,100 feet from the edge of the slide, in the vicinity of the old camp. Unlike the boulders in the slide, these are not solid, but have fresh fractures running through them, sometimes radiating from a central point; they did not travel by bounds, but fell from a height, remaining buried where they fell, and they are not found at other points around the edge of the slide.

A large quantity of dynamite belonging to the coal company was buried in the slide, but absolutely no trace of its explosion has been found anywhere.

Just west of the lower lake at the south end of the slide, a boulder clay terrace is partially buried under and partially cut away by the slide. The cutting appears to have been done by huge flying boulders, which shot through it. At one point a column of boulder clay is left standing alone.

The mouth of the upper tunnel of the mine is sealed with boulder clay, carried down from a deposit above and so compacted by pressure as to present the appearance of an original and undisturbed bed of boulder clay.

A few tree trunks, some of them scarcely bruised, were found mixed with the rock debris. Near the farthest limit of the slide a new tie, which must have come from the railway track, was found almost uninjured.

Dimensions

The dimensions of the slide cannot be given accurately, as no contoured maps showing the old topography are available for reference. A section along the central portion of the slide in the direction of the movement gave an average depth of 65 feet.

Figure 8
View looking down the valley across the slide-rock.
The uneven ridged surface of the slide and the varying sizes of the rock fragments are plainly seen. The contrast between the unaffected part of the valley and waste of slide-rock is also marked.

A section at right angles to the movement along the new location line of the railway, probably the thinnest portion of the slide, averages about 27 feet in depth. At one point north of the section measured the slide rock has a depth of fully 150 feet. At other points around the edge, the material is often spread out in a thin layer only a few feet in depth. The average depth of the whole of the slide rock, estimated from such measurements as we were able to obtain, amounts to about 45 feet. The area covered measures 1.03 square miles, or 3,190,528 square yards, and the contents of the mass, assuming 45 feet as the thickness, about 47,857,820 cubic yards. This represents approximately 36,000,000 cubic yards of solid rock, equal in weight to about 80,300,000 tons. A rough estimate of the size of the mass which broke away, obtained by comparison of the old and new contours of the mountain itself, is somewhat greater, amounting to about 40,702,000 cubic yards, or 90,796,000 tons.

Rock slides of this magnitude are exceedingly rare in the Canadian Rockies, in fact, no slide at all approaching it in size has been recorded from any part of the range. The mountains, as a rule, wear away gradually and fall in small blocks. A rock slide occurred in the Selkirk range, near Arrowhead, this spring. In the Alps destructive land slides have been more frequent, 150 having been recorded in Switzerland (Reidl, 1877). The dimensions of two of the largest of these, the Elm and Rossberg slides, may be inserted for comparison.

The Elm land slide (Heim, 1882; Rothpletz, 1881) was a 'bergsturz,' like that at Frank, only much smaller. The slide rock ran down a valley with a three per cent grade for a distance of 1,558 yards. Its width was 487 to 564 yards, its depth 16 to 21 yards. Its contents were estimated to be about 12,000,000 cubic yards. 84 houses and 115 human beings were buried beneath it.

The Rossberg slide in 1806, one of the largest on record, swept away four villages and killed 457 of their inhabitants. It was 4,300 yards long, 349 wide and 35 deep, and contained about 51,000,000 cubic yards of material (Baltzer, 1875; Zay, 1807).

The various theories first announced regarding the cause of the sudden destruction of a portion of the town of Frank, such as a volcanic outburst, a violent earthquake, an explosion in the coal mine, are, it is almost unnecessary to say, without foundation. None of the phenomena attendant on volcanic eruptions are observable. No earthquake shock is reported at the seismic stations, nor could an explosion in the mine, violent enough to satisfy the conditions have possibly occurred. The catastrophe was produced by a great land slide, ranking as one of the largest on record.

Ordinary atmospheric weathering, in which water plays the principal part, acting on rock masses, which on account of their material, form and structure are unstable, is the cause of by far the greatest number of rock slides. Jointing planes, or other fissures more or less strongly developed, are almost universal in the older rocks, and there are few which are not pervious to some extent to surface water. The water usually possesses some solvent power, and after entering the joint planes or fissures works its way downwards, gradually enlarging its channels and undermining the mass attacked. The increase of hydrostatic pressure of the water in the fissures by addition through rain or melting snow may be a sufficient cause of dislodgement of unstable masses of rock. The alternate freezing and thawing of such water-filled fissures in the rock is also a most efficient agent in loosening the material and, if the slope is sufficient, causing its fall. The expansion and contraction of the rocks themselves under rapid changes of temperature is not without effect. Earthquakes, or other violent shocks, often produce landslips even in fairly stable regions.

The second of the two principal classes of large land slides, mentioned in a previous chapter, the 'Bergrutsch,' is generally caused by water working its way along the bedding plane and softening a clayey layer, then, if the dip is sufficient, the upper part slips down along the lubricated plane. Many of the most disastrous land slips have been produced in this way.

The Frank slide was a 'Bergsturz,' a breaking away of the mountain mass across the bedding planes. Its primary cause is to be found in the structure and condition of Turtle mountain itself. It was ripe for a slide. The steep slopes, the shattered and fractured nature of the rocks, particularly of the basal beds of the limestone series, overlying the thrust-fault, coupled with unusually heavy precipitation are causes which in themselves are quite sufficient to have produced the slide and unaided the loosened masses would sooner or later have fallen. The pronounced earthquake movement of 1901, mentioned in Stupart's notes below, must have loosened the

Figure 9
Conical mound of debris on the surface of slide-rock and some of large boulders in the background which may have produced it.
The uneven surface of the slide-rock is apparent. A terrace, partially cut away, appears on the edge of the slide in the upper left hand corner.

threatening mass to some extent, thereby hastening the coming disaster. There are, however, two possible contributory causes, which may have fixed the time and date of the slide by breaking the last bands which held the unstable rock mass in place and may also have determined to some extent the dimensions of the present slide.

It is believed by many that the coal mine, situated at the base of the mountain, had something to do with the slide and a brief description of the workings is here given.

The mine, owned by the Canadian American Coal and Coke Company, is situated directly under the foot of the mountain. It is opened up by a tunnel 5,500 feet in length, commencing at the surface 45 feet above the bed of the river and running nearly southward along the vein approximately parallel to the face of Turtle mountain, and near the foot of the slope. The surface gradually rises in the direction of the tunnel, and at its end has a height above it of 1,190 feet (Figure 2). A second tunnel, used principally for air, situated 30 feet below the first, has been driven in for a distance of 2,560 feet, an upraise at 2,300 feet connects it with the upper tunnel. The coal seam, except where pierced by chutes, has been left intact for 32 feet above the main tunnel. The chambers mined are situated above this unworked strip, the larger ones commencing 1,200 feet from the mouth of the tunnel and continuing to 3,500 feet. They are divided into three sets, the first having a height of 250 feet, the second of 300 feet and the third of 400 feet. Beyond 3,500 feet, the chambers were just started, as is shown in the accompanying diagram (Figure 12). These chambers vary in length from 60 feet to 150 feet, but average about 130 feet and are separated by pillars usually 40 feet in length, containing man-ways and cross-cuts (Figure 13). All the workings are in coal. The seam is nearly vertical, dipping under the mountain at an angle of 82°, and has a width of thirteen to fifteen feet. Most of the coal stoped down in the chambers remained in them, just sufficient being drawn at the chutes to keep the surface of loose coal a convenient distance below the roof of unmined coal for the men to work.

Figure 10
Conical mound on surface of slide.

It is almost impossible to avoid the conclusion that these great chambers, 130 feet long, 250 to 400 feet high and 15 feet wide, situated directly under the foot of the mountain must have weakened it, even if, as the management assert, little of the loose coal had been drawn from them. The pressure on them must have been considerable. The loose coal, being less resistant than the unmined, would allow slight slips or readjustments in the hanging wall, and the jar produced by these may have been sufficient to snap some of the few remaining supports, which held the unbalanced mass in place. For in its state of unstable equilibrium, the slightest movement, a movement of even one inch, might have a profound effect upon Turtle mountain. It is a significant fact that the edges of the break correspond very closely with the limits of the big chambers and mined coal (Figure 12).

In this connection certain statements regarding the condition of the mine before the slide are of interest.

Statements Regarding the Condition of the Mine Before the Slide

The management state that the mine was in first class condition before the catastrophe, that the walls were solid, that there were no more movements or breaking away of the hanging wall than is usual in coal mines, and that the timbers were not under undue strain. Mr. Chestnut, a miner, states that slight movements were noticeable during the last seven months. These were particularly liable to occur between one and three in the morning. He describes them as like the starting and shuddering of a ship struck by a wave. Mr. Chapman also stated that these shocks were most frequent between the hours of one and three in the morning. These tremors were somewhat alarming to the miners, and some are said to have left the mine on account of them.

It is also reported that lately the coal has been mined with unusual ease, often running itself, so that the miners were taken off contract work and put upon day work. Rock from the hanging wall is said to have been falling in and mixing with the coal, so that men had to be employed in picking it out when the cars were dumped.

Cyrus Morris, formerly underground superintendent, stated that for the last seven months, there had been a general squeeze in the ground between 3,500 and 5,000 feet in the tunnel. The coal could be kept up only with difficulty. It was broken and would mine itself.

Small faults were encountered in the mine about every 120 feet, which had a slip of two feet into a hill, that is westward. At a thousand feet in, a fault was encountered with a throw of about fifteen feet in the same direction. Fissures extended into the hill at right angles to the strike of the seam, which brought in water, sulphur water and gas.

If the chambers were slightly overdrawn, leaving nine feet between loose coal and roof, the hanging wall would commence to break away along fractures and bedding planes, falling leaf after leaf, leaving a sort of tunnel in the wall sometimes over twelve feet long.

He was away for a few weeks preceding the slide, and during his absence he believes coal was run from some of the chambers, leaving them partially empty.

Mr. Frank B. Smith, B.Sc., M.E., Inspector of Mines, states that about 4,000 feet from the mouth of the tunnel a block from the hanging wall crushed in eighteen feet, making it necessary to run a tunnel around it. He also corroborates the evidence regarding tremors in the mine and the proneness of the hanging wall to break away.

Figure 11
View from the slope of Turtle Mountain, southward across the slide showing lakes formed at the base of the mountain by the ploughing out of the bed of Old Man River.
A hill of debris, 100 feet high, is seen just east of the lake shore.

Several witnesses stated that the shipments of coal were unusually heavy during the strike at Fernie, that is a little before the slide. As the force of men employed was not largely increased, some of the coal already mined must have been run from the chambers.

These statements are in accord with the evidence procured from an examination of the mountain in regard to the thoroughness with which the rocks are intersected by joints and fissures, and also with the conclusion regarding the pressure on the hanging walls, and the assumption that after mining the coal, readjustment would occur which would jar the mountain. They also indicate that these adjustments most frequently took place at a time (1 to 3 o'clock a.m.) when the rocks were contracting through cooling and when frost was most efficient in loosening them; that water was at work undermining the mass, and that the pressure on the hanging wall and seam was increasing as the threatening mass lost its supports one by one. If these statements are accepted, one must conclude that the working of the coal seam was connected to some extent at least with the dislodging of the central mass of Turtle mountain.

On the other hand, the testimony of the miners at work at the time of the disaster contains nothing that would indicate that the bursting of the last bond, by which the mass was upheld, was caused by movements in the mine. It indicates, rather, that anything which occurred in the mine was due to the slide. The slight difference in time noted by Chapman in the mine and that recorded outside seems to be without significance. Chapman's time agrees with that given by the fireman who was near the mouth of the mine, where he could observe only what was taking place outside. The mine appears to have escaped with little damage, much less than might be expected when the weight and force of the material which passed over it is taken into consideration. This slight effect is probably due to the shelf of rock above the seam, which caused the greater part of the rock mass to shoot over it into the valley. The mine, at time of writing, is still closed by debris, but when opened, it would be advisable to ascertain more definitely what movement, if any, has taken place.

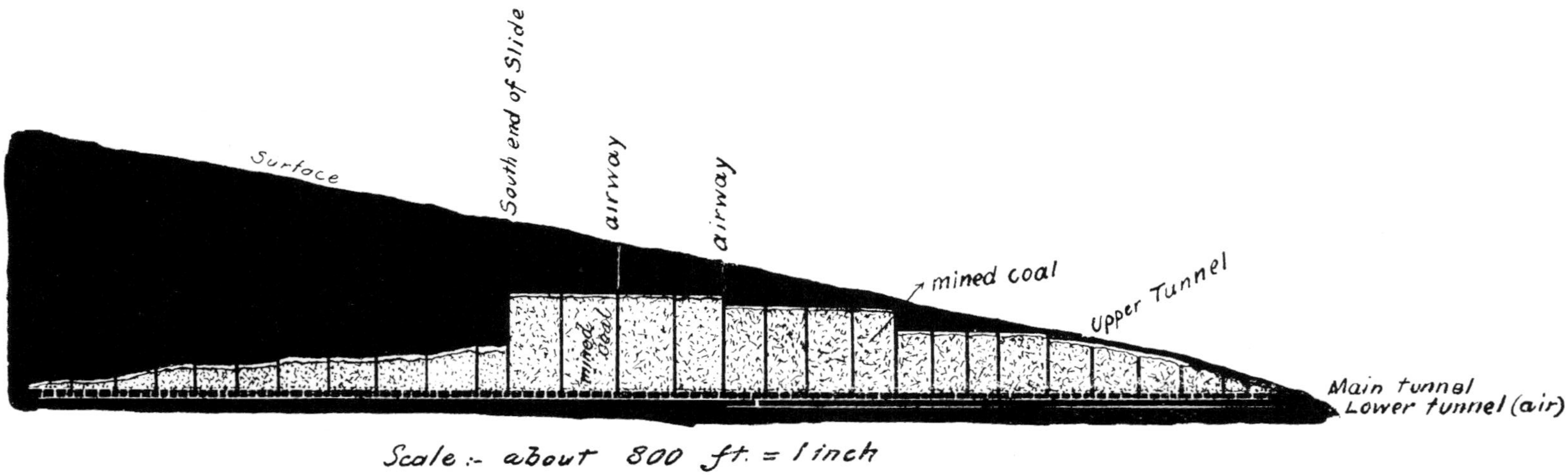

Figure 12
Diagram showing method of mining

The snapping of the last threads supporting the peak, which broke away, that is the final cause of the slide, was in all probability due to the temperature conditions during and preceding the morning of April 29.

The night of the slide was excessively cold. The miners say that it was colder than any night during the winter. Those outside state that the temperature was down to zero. The day before and the preceding days had been very hot, so that the fissures in the mountain must have been filled with water, on which the frost would act with powerful effect.

The rock slide cannot, therefore, be considered as due to a single cause, but rather, like so many phenomena in nature, to a combination of causes, cumulative in their effects. The chief of these were the structure and condition of the mountain, aided by exceptional atmospheric and other natural conditions, and also, possibly, by slight readjustments in the lower strata attendant on mining operations.

The following note on the meteorological and seismic conditions preceding the landslide has been kindly furnished by Mr. Stupart, Director of the Meteorological Service of Canada:—

"In compliance with the request contained in your letter of the 4th instant, I have much pleasure in replying to your questions in the order you have placed them.
'1. During several of the past few years the summer rainfall in Southern Alberta has been abnormally heavy. The average annual rainfall, exclusive of snow, at Calgary is 12.54 inches. In 1899 it was 21.61, 9.4 of which fell in the one month of August, 1900 was nearly average, 1901 was heavy, being 15.78 inches, 6.46 falling in June, and in 1902 it was phenomenal, as 28.90 inches fell, May 6.14 and also 28 inches of snow, June 8.82, July 5.06, August 8.14 inches.
'2. The precipitation so far this year has not been abnormal.
'3. An exceptionally warm wave passed over the country on the 24th and 25th April, on the latter day 74° was registered at Calgary.

This would have caused very rapid melting of the snow, with this exception, however, the months of March and April were both below average temperature.

'4. The total snowfall of this winter was less than average both at Pincher Creek and Calgary, although in both instances March was somewhat in excess.

'5. I do not know of any seismic disturbance of late years with centre of disturbance, to my knowledge, in that part of the country. The Toronto seismograph, however, registered a moderate disturbance six hours prior to the land slip at Frank. A most pronounced seismic disturbance was registered August 9th, 1901—a world shaking quake with centrum in the Aleutian Islands—large masses of ice were detached from Alaska glaciers, and I doubt not that the large blocks of ice in the Lynn Canal some days later, one of which wrecked the 'Islander,' were shaken off by this quake.

'6. There have been no severe electrical storms this spring."

Present Condition

The steep-walled summit of Turtle mountain was badly shattered and fractured by the shock received during the slide. The fractured portion has a width in places of 200 yards and extends from a point a couple of hundred feet north of the north peak, south to a point 400 feet south of the south peak, a distance of about half a mile.

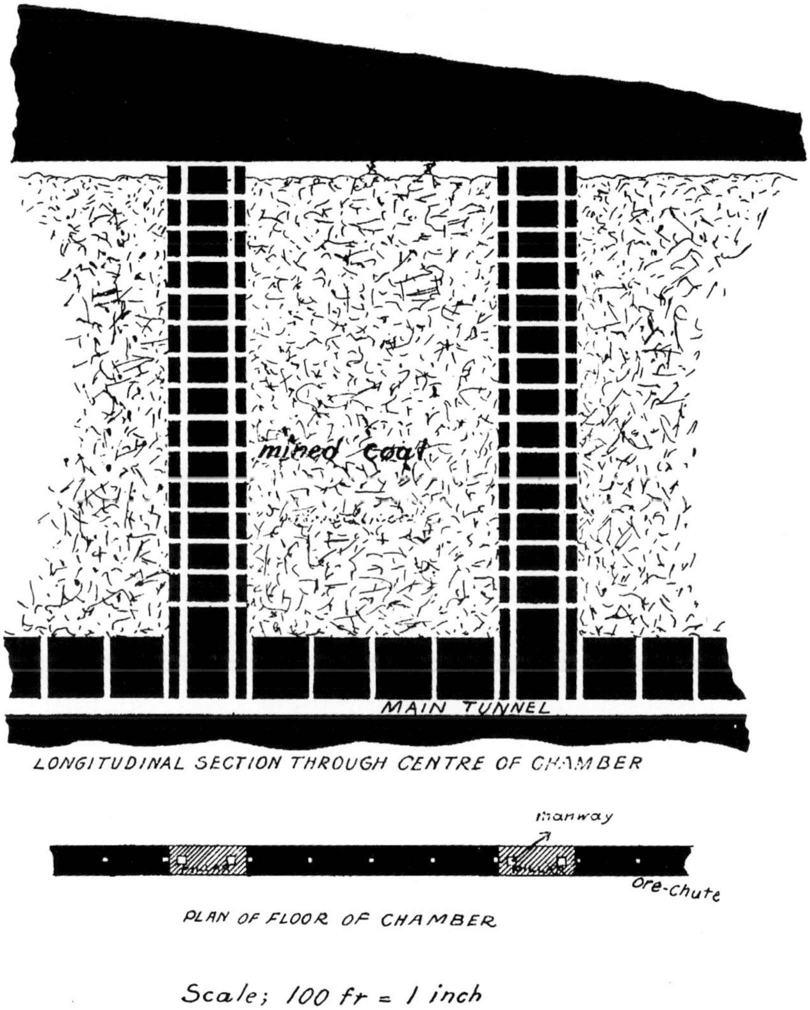

Figure 13
Longitudinal section through centre of chamber and plan of floor of chamber.

The fractures are too numerous to describe in detail, and only the more important ones will be referred to. A wide fissure running nearly parallel to the crest of the mountain cuts the western slope of the south peak, 100 yards west of the summit. The block of limestone between this fissure and the crest is broken by numerous smaller fissures and the whole shattered mass sunk forward fully 20 feet. Other fissures, affecting the stability of the south peak, were found cutting the eastern slope (Figure 14).

The strongest and most complicated fissuring occurs in the region between the north and south peaks, immediately behind the peak which fell away. The limestones here, for some distance west of the crest of the mountain, are cut by a net-work of gaping fissures into angular blocks, many of which are tilted forward 10° to 20° from their original position (Figure 14).

The fissuring in the north peak is less pronounced than in the south peak, but is important as this peak directly overhangs the town. The western slope of the peak is cut at intervals by narrow fissures, running roughly parallel to the face of the mountain for a distance of at least 500 feet west of the summit. One of these fissures, 150 feet from the summit slipped perceptibly during the six days' interval between our two examinations. The others, with the exception of a few short cracks near the face of the mountain, remained stationary. The small fissures north of the north peak also remained unmoved.

The fissures follow old jointing planes, often zigzagging from one set to another. Fresh fracture planes were seldom observed.

The shoulder of the mountain, running east from the north peak, and which looks down upon the town is excessively steep, having a slope for a considerable distance of 75°, and an average to the base of nearly 45°. To its stability Frank and its inhabitants owe their preservation. It withstood the concussion of the mountain mass breaking away, and acting like the side of a funnel prevented the loosened material from spreading over the town (Figure 15).

Figure 14
View from the north peak of Turtle Mountain to the south peak showing the edge of the break and the fissured zone behind.
From the angle at which the trees are bent forward from the vertical the degree of tilting of the blocks can be seen. Down the slope to the right of the left hand peak a large fracture can be distinguished by the break in the snow. Along this fracture the whole of this south peak has slipped at least 20 feet. The central peak, which was the highest, stood out to the left of the break about the middle of the picture.

The crushed limestone band above the thrust fault, which forms the base of one of the cliffs on this shoulder, has no great power of resistance. It is traversed by joints and slickens, which dip about 75° E., along which it readily weathers and falls away, leaving in places steep walled, fantastic 'hoodoos,' natural arches and pinnacles like earth pyramids, capped by remnants of a harder layer (Figure 5). The fractures, 525 feet behind the north peak, could not be followed northward to their termination on account of fresh snow and may extend far enough to loosen the whole of this shoulder of the mountain.

Present Dangers

The fractured zone surrounding the old break is bound sooner or later to fall away, but whether it falls gradually in small comparatively harmless blocks, or in large destructive masses depends upon future conditions, which cannot be foretold. The shattered mass between the north and south peaks does not menace the town, as the falling material will travel over the former slide. Moreover, the fracturing is so complete, that it is rapidly falling in individual blocks, which, though large, are yet not large enough for their momentum to carry them to the base of the mountain. During the six days between our visits, all the face which we had crossed between the north and south peaks, shown in Figure 14, had fallen away. This constant breaking away of the loosened blocks is decreasing the probability of further large rock slides.

That part of the north peak lying east of the fissure, 150 feet from the face, along which it slipped during our observations, threatens the mouth of the main tunnel of the mine, since, if it falls in one block, it will certainly reach the base of the mountain near the edge of the former slide. More danger to the town is apprehended from the fissures that exist behind and further to the west of the north peak. These fissures are narrow, but are wide enough to admit water, and the pressure of the rock above may keep them closed until the undermining action of the water, or some other cause liberates the superincumbent mass and a slide results. The safety of the town depends upon the stability of the shoulder protruding eastward from the north peak, which, as has been shown, is affected by them. The breaking away of the central portion of the mountain, which is going on continuously, is also tending to weaken this northern shoulder. If the town is to remain inhabited in its present position, these northern fissures must be closely watched. They are not likely to slip or extend suddenly (although, as stated above, there is always some liability of their doing so), and for this reason the upper portion of the town is not considered to be in any great immediate danger. If, however, any signs of slipping along the fissures some distance back from the north peak are detected, the town ought at once to be evacuated.

The town of Frank might exist on its present site uninjured for ages, but there will always be a possibility of a second destructive slide. The fact that the north shoulder withstood the shock of the first slide and was so solid that a snow cornice over its face was not broken down, is no proof that it is too solid to fall. Almost the same conditions exist on the north peak and shoulder to-day as obtained on the central peak before it broke away. A succession of seasons with unusually heavy precipitations and rapid changes of temperature, a slight earthquake shock, which is by no means an impossibility, or the closing of the chambers in the mine after the coal has been drawn, perhaps long after the inhabitants have lost all dread of the mountain, may snap the supports which retain this mass in place and start it on a career of destruction.

Since this possibility must always overhang the town it certainly seems in the interests of safety that it be moved a short distance up the valley, beyond the reach of danger.

As to the nature of the disaster at Frank there cannot be any doubt. It was a rock slide (Bergsturz) of colossal dimensions, one of the largest on record.

Turtle mountain, part of which fell away, is a sharp Rocky Mountain ridge, which with the exception of a short talus slope at the base, rises very steeply in a succession of cliffs to a height of 3,100 feet above the valley. It is the central and highest peak of this ridge which broke away. The mountain consists of Cretaceous sandstone or shales at the base, with limestone (part at least of Devonian age) forming all the upper portions. The limestone has been thrust over the newer rocks along a fault plane by the mountain building forces. The Cretaceous beds were overturned, dipping into the mountain at an angle of 82°, while the limestones have a westerly dip of 50° The basal beds of the limestones have been much folded and crumpled by friction, due to thrusting forward of the series along the fault plane. They form a very weak base for the precipitous mass above. The mountain is thoroughly dissected by fractures and jointing planes.

On the morning of April 29, a mass of rock surrounding the central peak, half a mile square and four or five hundred feet thick in the centre suddenly broke away and rushed with great velocity into and across the valley, destroying everything in its path. Fortunately the greater part of the town lay outside the course of the slide, but much property was destroyed and seventy lives were lost. Over a square mile of pleasant valley was transformed into a dismal waste of rock.

The motion of the slide rock was complex in detail, but as a whole resembled that of a viscous fluid. The distance from the top of the break to the foot of the slide, following the slope, is about two and a half miles. The average thickness of the slide rock across the valley, along the centre of the slide, is 65 feet, the deepest portion of the slide about 150 feet thick, and the average over the whole slide cannot be less than 45 feet. A rough estimate of the amount of material dislodged is about 40,000,000 cubic yards, representing in weight about 90,000,000 tons.

Figure 15
Shoulder projecting from the north peak and overhanging the town.
At the top right hand corner is the shattered north peak. From the lower left hand corner the shoulder drops for a thousand feet at an angle of 75°. The westward dip of the limestones is well shown. The mass which broke away lay just behind this shoulder, the break extending to the summit of the north peak at the top right hand corner of the figure.

In the Alps destructive land slides have been quite frequent, but are exceedingly uncommon in the Canadian Rockies. None, at all approaching this one in size has ever been recorded.

The slide was due not to a single cause, but to a combination of causes. The primary cause was undoubtedly the form and structure of Turtle mountain, already referred to. The huge mass was in a state of unstable equilibrium, possessed a weak base, and was thoroughly traversed by fissure and jointing planes, in which water and frost were continuously at work removing one by one the supports which held it in place. The heavy precipitation of the last few years accelerated this process. Recent earthquake tremors, particularly that of 1901, no doubt hastened the time of final disruption. The opening up of large chambers in the mine, situated under the base of the mountain, may have been a contributory cause, by allowing slight readjustments in the strata forming the hanging wall of the seam, producing jars that might dissever some of the few remaining bonds. The heavy frost on the morning of the slide, which followed hot summer-like days, appears to have been the force which severed the last thread and precipitated the unbalanced mass.

The zone surrounding the cavity left by the mass which broke away is excessively fractured and is doomed to fall. That portion behind the break and south of it is not dangerous, as it has a free course and is already falling in small individual blocks. The north peak and shoulder overhanging the town are, however, dangerous. They threaten the town and the mouth of the mine. Their condition now is very similar to that of the central peak before it fell away. There is always some possibility of a sudden movement, although the commencement of the movement is more likely to be gradual. In course of time, ordinary atmospheric weathering will bring down this threatening mass, but a slight earthquake shock, a succession of seasons with unusually heavy precipitation and rapid temperature changes, or the closing of the chambers in the mine, after the coal has been drawn, perhaps long after the inhabitants of the town have lost all dread of another disaster, may precipitate it suddenly in a second destructive slide.

Since this possibility must always overhang the town, it certainly seems advisable that it be moved a short distance up the valley, beyond the reach of danger.

Baltzer, A., 1875 — Über einen neuerlichen Felssturz am Rossberg, nebst einigen allgemeinen Bemerkungen über derartige Erscheinungen in den Alpen, (On a recent rockfall on Rossberg with a few general observations on these phenomena in the Alps), Neues Jahrbuch fur Mineralogie, Geologie und Paläontologie, pp. 15-26.

Baltzer, A., 1880 — Über Bergstürze, (On mountain collapses), Neues Jahrbuch fur Mineralogie, Geologie und Paläontologie, Band 2, pp. 197-199.

Heim, A., 1882 — Der Bergsturz von Elm, (The mountain collapse at Elm), Zeitschrift der Deutschen Geologischen Gesellschaft, Band 34, pp. 74-115.

McConnell, R.G., 1886 — Report on the geological structure of a portion of the Rocky Mountains accompanied by a section measured near the 51st Parallel, Geological and Natural History Survey of Canada, Annual Report, New Series, Volume 2, Part D, 41 p.

Reidl, E., 1887 — Über Bergsturz und Rutschung, (On mountain collapses and mountain ruptures), Neues Jahrbuch fur Mineralogie, Geologie und Paläontologie, pp. 914-924.

Rothpletz, A., 1881 — Der Bergsturz von Elm, (The mountain collapse at Elm), Zeitschrift der Deutschen Geologischen Gesellschaft, Band 33, pp. 540-564.

Willis, B., 1902 — Stratigraphy and structure, Lewis and Livingston Ranges, Montana, Bulletin. Geological Society of America, 13, pp. 305-352.

Zay, K., 1807 — Goldau und seine Gegend, (Goldau and its vicinity), Orell, Fuselli, Zürich.